Made in the USA
Columbia, SC
07 May 2022

60102635R00057